Welcome to the Excel Trick Course!

Excel is a powerful tool that is widely used by individuals and businesses alike. Whether you're a student, a financial analyst, a marketer, or a business owner, knowing how to use Excel effectively can greatly enhance your productivity and efficiency. However, with its multitude of features and functions, Excel can be overwhelming to navigate for beginners.

This course is designed to help you become an Excel power user by teaching you a variety of tricks and techniques that can save you time and effort. We will cover a range of topics, including shortcuts, formulas, data analysis, and visualizations, to name a few. By the end of the course, you will have the skills and knowledge to tackle complex tasks with ease and impress your colleagues with your Excel wizardry.

Whether you're new to Excel or a seasoned user looking to improve your skills, this course has something for everyone and let's get started on our Excel journey together!

Glossary:

Introduction

In today's fast-paced business world, technology plays a crucial role in decision-making, data analysis, and productivity. Microsoft Excel, one of the most widely used software applications in the world, has proven to be an indispensable tool for professionals across various industries. Excel has been around since the 1980s, and it continues to evolve with each new version, offering an increasing number of features and capabilities.

Excel has become the go-to tool for businesses of all sizes to manage their data, analyze it, and make decisions based on it. This software offers a range of functions and tools, including the ability to organize, sort, and filter data, create charts and graphs, and perform complex calculations. Excel is easy to use, and it has a low learning curve, making it an accessible tool for even the most inexperienced users.

Excel Use

Excel is used in various industries, including finance, accounting, marketing, and project management, among others. It is a crucial tool for financial analysts and accountants who use it to track and analyze financial data, create balance sheets, income statements, and financial forecasts. Excel is also widely used by marketing professionals to analyze market data, track sales trends, and create marketing plans.

Project managers use Excel to create and manage project schedules, track project costs, and monitor progress. In addition, Excel is a versatile tool for data scientists and researchers who use it to collect, analyze, and interpret data.

Excel formulas are powerful tools that allow users to perform calculations, manipulate data, and analyze information. Formulas can be used to perform a wide range of tasks, including adding, subtracting, multiplying, dividing, and more complex calculations such as statistical analysis or financial modeling.

Using Excel Formulas

To use Excel formulas, you first need to select the cell where you want to display the result of the calculation. Next, you need to type an equal sign (=) into the cell, which tells Excel that you are entering a formula. Then, you need to type in the formula itself. Excel formulas consist of a combination of values, cell references, and mathematical operators.

For example, to add two numbers, you can type "=2+2" into a cell. Excel will then display the result of the calculation in that cell, which in this case would be "4".

Excel also offers a range of built-in functions that can be used to perform more complex calculations. These functions can be accessed by typing in their names, followed by an open parenthesis "(", and the required arguments, separated by commas. For example, to calculate the average of a range of numbers, you can type "=AVERAGE(A1:A10)" into a cell. Excel will then calculate the average of the numbers in the range A1 to A10 and display the result in the cell.

Excel formulas can also be used to manipulate data, such as sorting and filtering data or creating customized reports. To sort data in Excel, you can use the SORT function, which allows you to sort data by column or row. To filter data, you can use the FILTER function, which enables you to filter data based on specific criteria.

Excel also offers a range of formulas that can be used for financial modeling, such as calculating interest rates, loan payments, and depreciation. These formulas can be used to create financial models that help businesses to make informed decisions about investments, loans, and other financial matters.

Excel formulas are powerful tools that allow users to perform calculations, manipulate data, and analyze information. By using formulas, users can automate tasks, save time, and increase accuracy. Excel offers a wide range of built-in functions that can be used to perform complex calculations, and these functions can be combined with mathematical operators, cell references, and other values to create powerful formulas. Whether you are a financial analyst, marketing professional, or project manager, Excel formulas are an essential tool for managing data and making informed decisions.

Excel Pivot Tables

Excel Pivot Tables are a powerful feature that allows users to summarize, analyze, and manipulate large datasets with ease. Pivot Tables enable users to convert rows of data into columns, group data by category, and create customized reports that summarize data in a meaningful way. This feature is particularly useful for users who need to analyze large datasets quickly and efficiently.

With Pivot Tables, users can easily sort and filter data, create calculated fields, and create visualizations that help them to understand the data better. This feature is particularly useful for financial analysts, accountants, and marketing professionals who need to analyze large datasets regularly.

Excel VBA

Excel VBA, or Visual Basic for Applications, is a programming language that enables users to create custom macros, automate repetitive tasks, and perform complex calculations. With VBA, users can create customized functions, add buttons to the Excel Ribbon, and automate tasks that would otherwise be time-consuming and repetitive.

VBA is particularly useful for users who work with large datasets and need to perform complex calculations regularly. By automating tasks and creating custom macros, users can save time and reduce the risk of errors.

Professional and Economic Importance of Excel

Excel has become a crucial tool for professionals across various industries, and it offers a range of benefits that can improve productivity, accuracy, and decision-making. Here are some of the reasons why Excel is so important:

Data Management: Excel offers a range of features that enable users to manage large datasets efficiently. Users can organize, sort, and filter data, create customized reports, and perform complex calculations.

Analysis: Excel enables users to analyze data quickly and efficiently, allowing them to identify trends, patterns, and outliers. This feature is particularly useful for financial analysts, marketing professionals, and data scientists who need to analyze large datasets regularly.

Visualization: Excel offers a range of charting and graphing tools that enable users to create visualizations that help them to understand data better. These visualizations are particularly useful for presentations and reports.

Automation: Excel VBA enables users to automate repetitive tasks, reducing the risk of errors and saving time. By automating tasks and creating custom macros, users can increase productivity and efficiency.

Cost-Effective: Excel is a cost-effective tool for managing data, analyzing it, and making decisions based on it. Compared to other software applications, Excel

Chapter 1: Formulas and data

Introduction:

Excel is a widely used spreadsheet software that offers a range of features and functionalities to its users. One of the key features that make Excel a popular tool for data analysis is its ability to perform various calculations and manipulations on data using formulas. Excel formulas allow users to automate complex calculations and analysis on large datasets, thus saving time and improving accuracy. In this chapter, we will discuss the 16 most commonly used Excel formulas, and provide a detailed explanation of how to use them over a dataset.

The total maximum size of lines in an Excel spreadsheet depends on the version of Excel you are using and the amount of memory available on your computer. In general, newer versions of Excel can handle larger datasets than older versions.

The maximum number of rows and columns in an Excel spreadsheet varies by version. In Excel 2003 and earlier versions, the maximum number of rows is 65,536, and the maximum number of columns is 256. In Excel 2007 and later versions, the maximum number of rows is over 1 million, and the maximum number of columns is 16,384.

However, the total maximum size of lines in an Excel spreadsheet also depends on the amount of memory available on your computer. If you are working with a large dataset, you may experience performance issues or crashes if you do not have enough memory to handle the amount of data you are working with.

In addition, there are other factors that can affect the maximum size of lines in an Excel spreadsheet, such as the complexity of formulas and calculations being performed, the number of charts and graphs in the spreadsheet, and the number of formatting options applied to the data.

To avoid performance issues and crashes when working with large datasets, it is important to optimize your Excel workbook by minimizing the number of formulas and calculations, using pivot tables and other features to summarize data, and removing unnecessary formatting and charting options. It may also be helpful to split large datasets into multiple smaller worksheets or workbooks to make them more manageable.

Using Excel Formulas over a Dataset:

The drag and drop feature is a useful tool when selecting data in Excel for use with formulas. This feature allows users to easily select data by clicking and dragging the mouse over a range of cells, rather than manually entering each cell reference into a formula.

Here are the steps for using the drag and drop feature for selecting data in Excel:

Step 1: Select the cell where you want to enter the formula.

Step 2: Type the equal sign (=) to start the formula.

Step 3: Click and hold the left mouse button on the cell you want to reference in the formula.

Step 4: Drag the mouse to highlight the range of cells you want to include in the formula.

Step 5: Release the mouse button when you have selected the desired range of cells.

Step 6: The cell references for the selected range of cells will automatically appear in the formula.

Step 7: Finish entering the formula and press Enter to calculate the result.

Using the drag and drop feature is a quick and efficient way to select large ranges of data in Excel, especially when working with large datasets. It can also help to reduce errors and typos that can occur when manually entering cell references into formulas. This feature is particularly useful when working with functions such as SUM, AVERAGE, COUNT, and others that require a range of cells as input.

Excel formulas can be used over a dataset to perform various types of calculations and analysis. The first step in using Excel formulas over a dataset is to enter the data into Excel. You can do this by uploading an Excel file, a pdf or even through a link from the web, Once the data is entered, it can be sorted, filtered, and formatted as required. Excel formulas can be used to analyze this data and extract meaningful insights.

The SUM formula is one of the most commonly used Excel formulas for analyzing datasets. It allows users to add up a range of values in a cell or multiple cells. For example, if we have a dataset of monthly sales figures for a particular product, we can use the SUM formula to calculate the total sales for the year. We simply need to select the range of cells containing the sales figures and enter the formula "=SUM(A1:A12)" in a new cell. This will add up the sales figures in cells A1 through A12, giving us the total sales for the year.

The AVERAGE formula is another commonly used Excel formula for analyzing datasets. It allows users to calculate the average of a range of values. For example, if we have a dataset of monthly temperature readings, we can use the

AVERAGE formula to calculate the average temperature for the year. We simply need to select the range of cells containing the temperature readings and enter the formula "=AVERAGE(A1:A12)" in a new cell. This will calculate the average temperature for the year based on the readings in cells A1 through A12.

The COUNT formula is useful for counting the number of cells that contain numbers in a range. For example, if we have a dataset of customer orders, we can use the COUNT formula to count the number of orders received in a particular month. We simply need to select the range of cells containing the order dates and enter the formula "=COUNTIF(A1:A12,"Jan")" in a new cell. This will count the number of orders received in January, based on the dates in cells A1 through A12.

The MAX and MIN formulas are useful for finding the highest and lowest values in a dataset. For example, if we have a dataset of stock prices, we can use the MAX formula to find the highest price and the MIN formula to find the lowest price. We simply need to select the range of cells containing the stock prices and enter the formula "=MAX(A1:A12)" or "=MIN(A1:A12)" in a new cell. This will give us the highest and lowest stock prices based on the values in cells A1 through A12.

The IF formula is useful for testing a condition and returning one value if the condition is true, and another value if the condition is false. For example, if we have a dataset of customer ratings for a product, we can use the IF formula to categorize the ratings as "good" or "bad". We simply need to select the range of cells containing the ratings and enter the formula "=IF(A1>3,"Good","Bad")" in a new cell. This will categorize the ratings as "Good" if they are greater than 3, and "Bad" if they are less than or equal to 3, based on the values in cells A1 through A12.

The top 16 most used Excel formulas

SUM: This formula adds up a range of values in a cell or multiple cells. For example, "=SUM(A1:A5)" will add up the values in cells A1 through A5.

AVERAGE: This formula calculates the average of a range of values. For example, "=AVERAGE(A1:A5)" will calculate the average of the values in cells A1 through A5.

COUNT: This formula counts the number of cells that contain numbers in a range. For example, "=COUNT(A1:A5)" will count the number of cells in cells A1 through A5 that contain numbers.

MAX: This formula returns the highest value in a range of cells. For example, "=MAX(A1:A5)" will return the highest value in cells A1 through A5.

MIN: This formula returns the lowest value in a range of cells. For example, "=MIN(A1:A5)" will return the lowest value in cells A1 through A5.

IF: This formula allows you to test a condition and return one value if the condition is true, and another value if the condition is false. For example, "=IF(A1>5,"Yes","No")" will return "Yes" if the value in cell A1 is greater than 5, and "No" if it is not.

SUMIF: This formula adds up a range of values based on a condition. For example, "=SUMIF(A1:A5,">5")" will add up the values in cells A1 through A5 that are greater than 5.

VLOOKUP: This formula searches for a value in the first column of a table and returns a corresponding value in the same row from another column. For example, "=VLOOKUP(A1,B1:C5,2,FALSE)" will search for the value in cell A1 in the first column of the table in cells B1 through C5, and return the value in the second column of the same row.

CONCATENATE: This formula joins two or more strings of text into one string. For example, "=CONCATENATE(A1," ",B1)" will join the values in cells A1 and B1 with a space in between.

LEFT: This formula extracts a specified number of characters from the left side of a text string. For example, "=LEFT(A1,3)" will extract the first three characters from the value in cell A1.

RIGHT: This formula extracts a specified number of characters from the right side of a text string. For example, "=RIGHT(A1,3)" will extract the last three characters from the value in cell A1.

MID: This formula extracts a specified number of characters from the middle of a text string. For example, "=MID(A1,3,5)" will extract five characters starting from the third character in the value in cell A1.

LEN: This formula returns the length of a text string. For example, "=LEN(A1)" will return the number of characters in the value in cell A1.

TRIM: This formula removes extra spaces from a text string. For example, "=TRIM(A1)" will remove any extra spaces from the value in cell A1.

SUBSTITUTE: This formula replaces a specified text string with another text string. For example, "=SUBSTITUTE(A1,"apple","orange")" will replace the word "apple" in the value in cell A1 with the word "orange".

ROUND: This formula rounds a number to a specified number of decimal places. For example, "=ROUND(A1,2)" will round the value in cell A1

The formula tab in the Excel main menu is an important tool for users who need to perform complex calculations and data analysis. The tab contains several options that allow users to create and manage formulas, functions, and other calculations within their Excel workbook. In this section, we will provide a detailed explanation of the different options available on the formula tab.

Formula Bar:

The formula bar is located at the top of the Excel window and displays the formula or function that is currently active in the selected cell. Users can directly edit the formula or function in the formula bar to make changes to their calculations.

Insert Function:

The insert function option allows users to search and select from a list of predefined functions that can be used in their calculations. Users can browse functions by category or use the search bar to find a specific function.

Name Manager:

The name manager option allows users to manage named ranges and formulas within their Excel workbook. Users can create, edit, and delete named ranges and formulas, as well as assign names to specific cells or ranges.

Define Name:

The define name option allows users to create and manage named ranges and formulas. Users can assign a name to a specific cell or range of cells, and use the named range in their calculations.

Create from Selection:

The create from selection option allows users to create a named range based on the selected cells in their worksheet. Users can specify the name and scope of the named range, and use it in their calculations.

Evaluate Formula:

The evaluate formula option allows users to step through a formula and see how each part of the formula is calculated. Users can use this option to debug complex formulas and functions and ensure their calculations are accurate.

Trace Precedents:

The trace precedents option allows users to visually identify the cells that are referenced in a selected cell's formula. Users can use this option to trace the flow of data in their calculations and ensure the accuracy of their formulas.

Trace Dependents:

The trace dependents option allows users to visually identify the cells that depend on the selected cell for their calculations. Users can use this option to trace the impact of changing a particular cell on the rest of their calculations.

Remove Arrows:

The remove arrows option allows users to remove the visual indicators of precedents and dependents from their worksheet. Users can use this option to clean up their worksheet and remove clutter from their calculations.

Error Checking:

The error checking option allows users to identify and correct errors in their formulas and functions. Users can use this option to check for common errors, such as circular references, and fix them before they impact their calculations.

In conclusion, the formula tab in the Excel main menu is a powerful tool for users who need to perform complex calculations and data analysis. The tab contains several options that allow users to create and manage formulas, functions, and other calculations within their Excel workbook. The different options available on the formula tab include the formula bar, insert function, name manager, define name, create from selection, evaluate formula, trace precedents, trace dependents, remove arrows, and error checking. By using these options effectively, users can ensure the accuracy and reliability of their calculations and achieve their data analysis goals more efficiently.

Uploading data into Excel is a key task for users who need to analyze and manipulate data. There are several methods to upload data into Excel, including using an Excel file, a PDF, a CSV file, and using a link. In this section, we will provide a detailed explanation of each method.

icrosoft Excel is a powerful tool for data analysis, and one of its greatest strengths is its vast library of pre-made formulas. In the latest version of Excel, there are hundreds of built-in formulas that can help you to perform complex calculations, manipulate data, and automate common tasks.

These formulas can be found in the Formulas tab in the main menu of Excel. From here, you can browse and search through the different categories of formulas, such as Math & Trig, Date & Time, Lookup & Reference, and more. Each formula has its own syntax and set of arguments, which you can customize to fit your specific needs.

Some of the most commonly used Excel formulas include SUM, AVERAGE, COUNT, MAX, MIN, and IF. These formulas are useful for performing basic arithmetic calculations, counting values in a dataset, and testing conditions.

However, there are many more advanced formulas available in Excel that can be used to handle larger and more complex datasets. For example, the SUMIF and COUNTIF formulas can be used to sum or count values in a dataset based on certain criteria. The VLOOKUP and HLOOKUP formulas can be used to search for and retrieve data from a table, while the INDEX and MATCH formulas can be used to look up data in a table based on multiple criteria.

In addition to these pre-made formulas, Excel also offers the ability to create custom formulas using VBA (Visual Basic for Applications). VBA is a programming language that is built into Excel, and it allows you to create your own custom functions and macros to automate repetitive tasks and perform more advanced data analysis.

In later chapters, we will cover the use of Excel VBA for applications, including how to create your own custom formulas and macros. This will enable you to handle even larger datasets and perform more complex data analysis tasks, helping you to become a more skilled and efficient Excel user.

Other advanced use formulas and examples of how to use them:
how to use the COUNTIF, MATCH, INDEX, and HLOOKUP functions in Excel.

COUNTIF Function

The COUNTIF function is used to count the number of cells in a range that meet a certain criteria. To use this function, you first select the range of cells you want to count, and then specify the criteria in the formula.

For example, let's say you have a list of sales figures for different products, and you want to count the number of products that have sales of more than $1000. You can use the COUNTIF function with the following formula:

=COUNTIF(B2:B10,">1000")

This formula will count the number of cells in the range B2:B10 that are greater than 1000.

MATCH Function

The MATCH function is used to find the position of a value in a range. To use this function, you specify the value you want to find, and the range of cells you want to search.

For example, let's say you have a list of employee names and their salaries, and you want to find the position of an employee with a specific salary. You can use the MATCH function with the following formula:

=MATCH(40000,B2:B10,0)

This formula will search the range B2:B10 for the value 40000, and return the position of the first cell that contains that value.

INDEX Function

The INDEX function is used to retrieve a value from a specific row and column in a table. To use this function, you specify the table range, the row number, and the column number.

For example, let's say you have a table of sales data for different products and months, and you want to retrieve the sales figure for a specific product and month. You can use the INDEX function with the following formula:

=INDEX(B2:F10,3,4)

This formula will retrieve the value from the third row and fourth column of the range B2:F10, which corresponds to the sales figure for the third product in the fourth month.

HLOOKUP Function

The HLOOKUP function is used to search for a value in the top row of a table, and return the corresponding value from a specified row in the table. To use this function, you specify the lookup value, the table range, the row number, and the range lookup option.

For example, let's say you have a table of sales data for different products and months, and you want to retrieve the sales figure for a specific product in a specific month. You can use the HLOOKUP function with the following formula:

=HLOOKUP("Product C",B1:F10,3,FALSE)

This formula will search the top row of the range B1:F10 for the value "Product C", and return the corresponding value from the third row of the table, which corresponds to the sales figure for Product C in the specified month.

In summary, these functions are powerful tools in Excel that can help you to manipulate and analyze large datasets. By understanding how to use these functions, you can become a more efficient and effective Excel user.

Uploading Data from an Excel File:

Uploading data from an Excel file is a straightforward process. First, open a new Excel workbook and navigate to the "Data" tab. From there, click on "From File" and select "From Excel". Then, navigate to the location where the Excel file is saved and select it. If the file is password-protected, you will need to enter the password to access the data.

Once you have selected the Excel file, Excel will prompt you to select the range of data you want to import. You can choose to import the entire workbook or select specific sheets and ranges. After selecting the data you want to import, click "OK" to upload the data into Excel.

Uploading Data from a PDF:

Uploading data from a PDF file can be a bit more challenging than uploading from an Excel file. First, you will need to convert the PDF file into an Excel-readable format, such as a CSV or XLSX file. There are several online tools available for converting PDF files to Excel, or you can use a dedicated PDF conversion software.

Once you have converted the PDF file into an Excel-readable format, you can follow the same process as uploading data from an Excel file. Open a new Excel workbook and navigate to the "Data" tab. From there, click on "From File" and select "From Text/CSV" if the converted file is a CSV file, or select "From Excel" if the converted file is an XLSX file. Then, navigate to the location where the converted file is saved and select it.

Excel will prompt you to select the range of data you want to import, and you can choose to import the entire file or specific sheets and ranges.

Uploading Data from a CSV File:

Uploading data from a CSV file is similar to uploading from an Excel file. First, open a new Excel workbook and navigate to the "Data" tab. From there, click on "From File" and select "From Text/CSV". Then, navigate to the location where the CSV file is saved and select it. Excel will prompt you to select the range of data you want to import, and you can choose to import the entire file or specific sheets and ranges.

Uploading Data from a Link:

Uploading data from a link is a convenient way to access and analyze data without needing to download a file. To upload data from a link, open a new Excel workbook and navigate to the "Data" tab. From there, click on "From Web" and enter the URL of the webpage containing the data you want to import. Excel will open a web query window and allow you to select the data you want to import.

Once you have selected the data you want to import, click "Import" to upload the data into Excel. Excel will prompt you to select the location where you want to save the imported data, and you can choose to save it as a new worksheet or overwrite an existing one.

In conclusion, uploading data into Excel can be done through several methods such as using an Excel file, a PDF, a CSV file, and using a link. The process for each method varies slightly, but the overall steps involve navigating to the location where the data is stored, selecting the data range to import, and saving the imported data in the desired location within Excel.

Mockaroo.com and Mockupdata.com are online tools that allow users to generate mock data for use in Excel and other applications. They are useful for Excel users who want to practice their knowledge and skills with a variety of data types and formats.

One of the main advantages of using these tools is that they allow users to generate large datasets quickly and easily. This is especially useful for users who want to practice their skills with a variety of data types and formats, but do not have access to large datasets for testing and experimentation. With these tools, users can generate datasets of any size and complexity, and can customize them to meet their specific needs.

Mockaroo.com and Mockupdata.com are also useful for Excel users who want to practice their knowledge and skills with new Excel functions and formulas. By generating mock data with these tools, users can practice their skills with a variety of functions and formulas, and can experiment with different techniques for working with large datasets.

In addition, these tools are useful for testing and development purposes, as well as for creating realistic datasets for training and educational purposes. They can be used to create test datasets for software applications, to create training materials for Excel users, and to create sample datasets for research and analysis purposes.

Overall, Mockaroo.com and Mockupdata.com are valuable resources for Excel users who want to practice their skills with a variety of data types and formats. Whether you are a beginner or an advanced Excel user, these tools can help you improve your knowledge and skills, and can help you become more proficient at working with large datasets in Excel.

Saving and managing data files is an essential aspect of working with Microsoft Excel. It is important to save your work frequently to avoid losing any changes you have made to your data. Additionally, it is important to manage your data files properly to ensure that they are organized and easy to find when you need them.

Here are some reasons why saving and managing data files is important:

Avoid Data Loss: By saving your work frequently, you can avoid losing any changes you have made to your data. If you experience a power outage, computer crash, or other unexpected interruption, you may lose any unsaved work. By saving your work frequently, you can minimize the risk of losing your data.

Improve Efficiency: When you save and manage your data files properly, you can easily find the files you need when you need them. This can save you time and improve your overall efficiency when working with Microsoft Excel.

Protect Your Data: It is important to protect your data from unauthorized access, theft, and other security threats. By managing your data files properly, you can ensure that your data is stored securely and is only accessible to authorized users.

Maintain Version Control: When working with data files, it is important to maintain version control to ensure that you are always working with the most up-to-date version of your data. By managing your data files properly, you can ensure that you are always working with the latest version of your data.

Reduce Errors: When you manage your data files properly, you can reduce the risk of errors and inconsistencies in your data. By following an organized and consistent process for working with data files, you can ensure that your data is accurate and reliable.

To ensure that you are saving and managing your data files properly, you should follow these best practices:

Save your work frequently: Save your work every few minutes to ensure that you don't lose any changes you have made to your data.

Use descriptive file names: Use descriptive file names that reflect the content of the file to make it easier to find the file when you need it.

Organize your files into folders: Organize your files into folders based on their content or purpose to make it easier to find them when you need them.

Use version control: Use version control to ensure that you are always working with the latest version of your data.

Make backups: Make regular backups of your data files to protect them from loss or damage.

By following these best practices, you can ensure that you are saving and managing your data files properly, and can avoid data loss, improve efficiency, protect your data, maintain version control, and reduce errors.

The filter feature in Microsoft Excel allows you to sort and display specific data based on your selected criteria. Here are the steps to use the filter in Excel:

Select the range of cells that you want to filter.

Go to the Data tab on the Ribbon at the top of Excel.

Click on the Filter button in the Sort & Filter group. This will add filter arrows to the top row of your selected range.

Click on the filter arrow for the column that you want to filter by. This will open a drop-down menu.

In the drop-down menu, you can select from various filter options such as text filters, number filters, and date filters. You can also use search or sort options to find specific data.

Select the filter criteria that you want to apply to the column. For example, if you want to display all records that contain the word "Apple," you can select the "Text Filters" option and enter "Apple" in the search box.

Click OK to apply the filter to the column. You can also filter by multiple criteria by selecting the "Filter by Color" or "Custom Filter" options.

To remove the filter, simply click on the filter arrow again and select "Clear Filter."

Overall, using the filter in Microsoft Excel can help you quickly and easily sort through large amounts of data to find the information you need.

Chapter 2: Pivot tables

Microsoft Excel is a powerful tool for data analysis and management. One of the most valuable features of Excel is the PivotTable, which allows users to easily summarize and analyze large amounts of data in a dynamic and interactive way. In this guide, we will explore how to use PivotTables in Microsoft Excel to create meaningful summaries of data, identify trends, and gain insights into your data.

What is a PivotTable?

A PivotTable is a powerful data analysis tool in Excel that allows you to summarize and analyze large amounts of data in a flexible and interactive way. PivotTables allow you to quickly and easily organize, filter, and sort data, as well as create meaningful summaries and visualizations of your data. PivotTables are particularly useful when working with large datasets that require more complex analysis and visualization.

PivotTables work by allowing users to "pivot" or "rotate" their data, which means reorganizing the data so that it can be easily summarized and analyzed in different ways. PivotTables allow users to group and aggregate data by different categories, such as dates, regions, products, or customers, and to apply different calculations, such as sums, averages, and counts, to those groups. PivotTables also allow users to filter and sort data, and to create visualizations such as charts and graphs.

How to Create a PivotTable

Creating a PivotTable in Excel is a simple and straightforward process. Here are the steps to create a PivotTable:

Step 1: Select the range of data that you want to analyze. This range should include column headings and row labels.

Step 2: Click the "Insert" tab in the Excel ribbon, and select "PivotTable" from the dropdown menu.

Step 3: In the "Create PivotTable" dialog box, select the range of data that you want to analyze, and choose where to place the PivotTable (in a new worksheet or an existing one).

Step 4: Click "OK" to create the PivotTable.

Step 5: The PivotTable Field List will appear on the right side of the screen. This is where you will choose which data to include in the PivotTable.

Step 6: Drag and drop the column headings and row labels into the appropriate areas in the PivotTable Field List. The column headings should be placed in the "Columns" area, the row labels should be placed in the "Rows" area, and any data you want to summarize should be placed in the "Values" area.

Step 7: Choose the appropriate calculation to apply to the data in the "Values" area. This can be a sum, average, count, or any other calculation that you want to apply.

Step 8: Customize the PivotTable by filtering, sorting, or formatting the data as needed.

Step 9: Refresh the PivotTable if the source data changes.

Using PivotTables for Data Analysis

Pivot tables are incredibly useful for data analysis and management in Microsoft Excel. Here are some of the key utilities of pivot tables:

Data Summarization: Pivot tables allow you to summarize large amounts of data in a dynamic and interactive way. You can easily group data by different categories, such as dates, regions, products, or customers, and apply different calculations, such as sums, averages, and counts, to those groups. This helps to create meaningful summaries and visualizations of your data, making it easier to analyze and interpret.

Data Filtering: Pivot tables allow you to filter data, so that you can focus on specific subsets of your data. For example, you can filter your data to show only sales data for a specific region, or to show the top 10 products by sales. This makes it easy to drill down into your data and uncover insights that might not be apparent at first glance.

Data Sorting: Pivot tables allow you to sort your data in different ways, such as by date, product, or customer name. This helps to identify patterns and trends in your data, making it easier to understand and analyze.

Data Visualization: Pivot tables allow you to create visualizations, such as charts and graphs, that help to convey complex data in a clear and concise way. This helps to identify trends and patterns in your data, making it easier to draw insights and conclusions.

Data Comparison: Pivot tables allow you to compare data across different categories, such as comparing sales data across different regions or products. This helps to identify areas where performance is strong or weak, and to identify opportunities for improvement.

Data Forecasting: Pivot tables allow you to forecast future trends based on past data. This is particularly useful for businesses that need to plan ahead and make decisions based on future projections.

Overall, pivot tables are incredibly useful for data analysis and management, and are a key tool for businesses and individuals who need to work with large amounts of data. By allowing you to quickly and easily organize, filter, and summarize data, pivot tables can help you to uncover insights and make informed decisions based on your data.

Chapter 3: Excel features

Microsoft Excel is a powerful spreadsheet software that offers a wide range of features and tools to help users work more efficiently and effectively with their data. Among the many features available in Excel, the Page Layout, Grammar Spelling Check, and View tabs are some of the most useful and frequently used.

Page Layout Tab

The Page Layout tab in Excel is used to manage the appearance of the worksheet, such as margins, page orientation, and print settings. It provides a variety of tools and options to help users prepare their worksheets for printing, as well as to make them more visually appealing and organized.

One of the most useful tools in the Page Layout tab is the Page Setup dialog box, which allows users to adjust a variety of settings related to the layout and formatting of their worksheets. For example, users can set the page orientation to either portrait or landscape, adjust the margins, and specify the print area. Additionally, the Page Layout tab also includes tools for adding headers and footers, adjusting the scaling of the worksheet, and adding page breaks.

The Page Layout view in Excel is a powerful tool for managing the layout and appearance of a worksheet, especially when preparing it for printing. By using this view, you can ensure that your worksheet looks exactly the way you want it to when it's printed.

Here are the steps to use the Page Layout view to print an Excel sheet correctly:

Click on the View tab at the top of the Excel window.

Click on the Page Layout button in the Workbook Views section.

This will change the view to the Page Layout view, which shows the worksheet as it will appear when printed.

Use the tools in the Page Layout tab to adjust the layout and appearance of the worksheet. This may include setting margins, adjusting the page orientation, and adding headers and footers.

Once you have made all the necessary adjustments, preview the worksheet in print preview mode by clicking on the File tab and selecting Print.

In the Print Preview window, you can see exactly how the worksheet will appear when printed.

Use the print settings to adjust the printer and page settings as necessary. This may include selecting the printer, setting the number of copies, and adjusting the print quality.

Click the Print button to print the worksheet.

By using the Page Layout view and following these steps, you can ensure that your worksheet is printed exactly the way you want it to look. This can be particularly helpful when working with large datasets or when preparing reports or other documents that require a professional appearance.

Grammar Spelling Check Tab

The Grammar Spelling Check tab in Excel is used to check the spelling and grammar of the content in the worksheet. It provides users with a variety of options for checking their spelling and grammar, as well as correcting any errors that are found.

One of the most useful tools in the Grammar Spelling Check tab is the Spelling dialog box, which can be used to check the spelling of individual words, as well as to correct any errors that are found. Additionally, the Grammar Spelling Check tab also includes tools for checking the grammar of the content in the worksheet, as well as for customizing the settings for the spelling and grammar check.

View Tab

The View tab in Excel is used to manage the way that the worksheet is displayed on the screen. It provides a variety of options for adjusting the view of the worksheet, such as zooming in or out, splitting the screen, and changing the view mode.

One of the most useful tools in the View tab is the Zoom tool, which allows users to adjust the size of the worksheet on the screen. This is particularly useful when working with large datasets or when trying to view specific details in the worksheet. Additionally, the View tab also includes tools for splitting the screen into multiple panes, as well as for changing the view mode to better suit the user's needs.

In conclusion, the Page Layout, Grammar Spelling Check, and View tabs in Excel are important features that offer a wide range of tools and options to help users work more efficiently and effectively with their data. By understanding how to use these features, users can better manage the appearance and content of their worksheets, as well as improve the accuracy and clarity of their work.

Chapter 4: Graphics

To select data in Excel for use in graphs and charts, follow these steps:

Open the Excel spreadsheet containing the data you want to use in your graph.

Highlight the cells containing the data you want to use. You can click and drag to select a range of cells, or you can hold down the "Ctrl" key on your keyboard and click on individual cells to select them.

Once you have selected your data, click on the "Insert" tab in the Excel ribbon.

In the "Charts" section of the ribbon, you will see various types of charts and graphs that you can create. Click on the type of chart you want to create, such as a bar chart, line chart, or pie chart.

Excel will automatically generate a chart based on the data you have selected. You can then customize the chart by adding titles, labels, and other formatting options using the tools in the "Chart Design" and "Chart Format" tabs that appear when the chart is selected.

If you want to change the data used in the chart, simply click on the chart to select it, then click on the "Select Data" button in the "Data" section of the ribbon. From here, you can add or remove data series or change the range of cells used for the chart.

Once you have finished customizing your chart, you can either keep it on the same worksheet as your data or move it to a separate worksheet by selecting the chart and then clicking on the "Move Chart" button in the "Location" section of the "Chart Design" tab.

The Insert tab in Microsoft Excel provides a variety of tools and options for adding and manipulating different types of content within a worksheet. In this section, we will discuss each of the options available in the Insert tab in detail.

Tables: The Tables option in the Insert tab allows you to quickly convert a range of cells into a table. Tables provide a range of built-in functionality, including sorting, filtering, and formatting options. Simply select the range of cells you want to convert to a table, click the Tables button, and select the desired table style.

Charts: The Charts option in the Insert tab allows you to create a variety of different chart types, including column, line, pie, bar, area, and more. To create a chart, select the data you want to include in the chart, click the Charts button, and select the desired chart type.

PivotTables: The PivotTables option in the Insert tab allows you to create a powerful tool for analyzing and summarizing large datasets. PivotTables allow you to quickly and easily summarize data by category, and can be used to create charts and other visualizations. To create a PivotTable, select the data you want to summarize, click the PivotTables button, and follow the prompts.

Sparklines: The Sparklines option in the Insert tab allows you to create small, simple charts that can be inserted directly into cells. Sparklines are ideal for visualizing trends and patterns in data, and can be easily customized to fit your needs. To create a sparkline, select the range of cells you want to include in the chart, click the Sparklines button, and select the desired sparkline type.

Pictures: The Pictures option in the Insert tab allows you to add images and graphics to your worksheet. Simply click the Pictures button, browse to the location of the image file, and insert it into the worksheet.

Shapes: The Shapes option in the Insert tab allows you to add a variety of different shapes to your worksheet, including lines, arrows, rectangles, and circles. Simply select the desired shape from the Shapes dropdown, draw the shape on the worksheet, and customize it as needed.

SmartArt: The SmartArt option in the Insert tab allows you to create professional-looking diagrams and charts. SmartArt is a great way to visualize complex information, and can be easily customized to fit your needs. To create SmartArt, click the SmartArt button, select the desired layout, and enter your content.

Slicers: The Slicers option in the Insert tab allows you to quickly filter and analyze large datasets. Slicers allow you to easily create interactive dashboards and reports, and can be used to filter data by category or value. To create a slicer, select the data you want to filter, click the Slicers button, and select the desired slicer type.

Headers & Footers: The Headers & Footers option in the Insert tab allows you to add custom headers and footers to your worksheet. Headers and footers can be used to add important information to your worksheet, such as page numbers, document titles, and dates. To add a header or footer, click the Header & Footer button, select the desired option, and enter your content.

Text: The Text option in the Insert tab allows you to add text boxes, WordArt, and other text-based elements to your worksheet. To add text, click the Text button, select the desired option, and enter your content.

Overall, the Insert tab in Microsoft Excel provides a wide range of tools and options for adding and manipulating different types of content within a worksheet. By taking advantage of these options, you can create professional-looking reports, visualizations, and dashboards that are both informative and professional.

Line and column charts are two of the most commonly used chart types in Excel. Here's how to create and customize them:

Creating a Line Chart in Excel:

Select the range of data that you want to use for the line chart.

Click on the "Insert" tab in the Excel ribbon.

In the "Charts" section of the ribbon, click on the "Line" button to see the different line chart options.

Select the type of line chart you want to create, such as a basic line chart or a stacked line chart.

Excel will create the line chart based on your data. You can then customize the chart by adding titles, labels, and other formatting options using the tools in the "Chart Design" and "Chart Format" tabs that appear when the chart is selected.

Creating a Column Chart in Excel:

Select the range of data that you want to use for the column chart.

Click on the "Insert" tab in the Excel ribbon.

In the "Charts" section of the ribbon, click on the "Column" button to see the different column chart options.

Select the type of column chart you want to create, such as a clustered column chart or a stacked column chart.

Excel will create the column chart based on your data. You can then customize the chart by adding titles, labels, and other formatting options using the tools in the "Chart Design" and "Chart Format" tabs that appear when the chart is selected.

Customizing Your Charts:

Once you have created your line or column chart, you can customize it further using the following options:

Chart Titles: Add a title to your chart by clicking on the "Chart Title" button in the "Chart Layouts" section of the "Chart Design" tab.

Axis Labels: Add labels to the x and y axis of your chart by clicking on the "Axis Titles" button in the "Chart Layouts" section of the "Chart Design" tab.

Data Labels: Show the data labels on your chart by clicking on the "Data Labels" button in the "Labels" section of the "Chart Design" tab.

Legends: Show or hide the legend on your chart by clicking on the "Legend" button in the "Labels" section of the "Chart Design" tab.

Formatting: Customize the look and feel of your chart by using the formatting options in the "Chart Format" tab.

By following these steps, you can create and customize line and column charts in Microsoft Excel to display your data in an easy-to-understand format.

Chapter 5: Introduction to macros

VBA (Visual Basic for Applications) macros are a powerful tool that allows users to automate tasks in Excel. Macros are essentially a series of commands that are written in the VBA programming language and can be used to manipulate data, automate tasks, and create custom functions within Excel.

One of the key benefits of using VBA macros in Excel is their ability to handle large datasets. With macros, you can quickly and easily perform complex calculations on thousands of rows of data, without the need for manual data entry or manipulation.

In addition to handling large datasets, VBA macros also allow you to manipulate data in a variety of ways, including sorting, filtering, and transforming data into different formats. You can also use macros to create custom reports and dashboards that allow you to quickly analyze and visualize your data.

Another key feature of VBA macros is their ability to create user forms, which can be used to collect data from users and automate data entry. User forms can be customized with various input fields, such as drop-down lists, checkboxes, and text boxes, making it easy to gather and process data from users.

Finally, VBA macros can also be used to create custom Excel functions, which can be used to perform specific calculations or manipulate data in a particular way. Custom functions can be used in the same way as standard Excel functions, and can be saved and reused across multiple workbooks.

Overall, VBA macros play a crucial role in the handling of large datasets in Excel, allowing users to automate tasks, manipulate data, and create custom functions to meet their specific needs. By leveraging the power of VBA macros, users can save time and effort, while improving the accuracy and efficiency of their data analysis and reporting processes.

In order to be able to use macros, customized user forms to work with data, customized new functions or for example to manage the format of a very large file you will need to actívate macros on your licensed Microsoft Excel, state that this will only work with the licensed desktop versión of Excel and it is not available on the free online version available at Microsoft.com

Check here the instructions to actívate macros:

By default, macros are disabled in Excel for security reasons. To use macros in Excel, you need to enable them by changing the macro settings. Here's how you can do it:

Open Excel and click on the "File" tab in the top-left corner of the screen.

Click on "Options" in the left-hand menu.

In the Excel Options dialog box, click on "Trust Center" in the left-hand menu.

Click on the "Trust Center Settings" button on the right-hand side of the dialog box.

In the Trust Center dialog box, click on "Macro Settings" in the left-hand menu.

Choose one of the following options:

"Disable all macros without notification": This option will disable all macros without warning you, and you won't be able to use macros in Excel.

"Disable all macros with notification": This option will disable all macros by default, but will prompt you with a warning message whenever you open a workbook that contains macros. You can choose to enable the macros if you trust the source of the workbook.

"Enable all macros": This option will enable all macros without prompting you, and you can use macros in Excel without any warnings.

After you've selected the macro settings, click on "OK" to close the dialog boxes.

Once you've enabled macros in Excel, you can use them by opening a workbook that contains macros and following the instructions provided by the macro. Note that it's important to be cautious when using macros, especially if you're not familiar with the source of the workbook. Always make sure that you trust the source of the workbook before enabling macros.

After activating macros you will be able to Access the embded visual studio code editor for Microsoft Excel by pressing the keys Alt plus F11 at the same time.

In this editor you will find some blank code pages for the sheet you are currently working and on the left side of our screen you will aswell find the code file for the module, the sheet in where you will need to place your code, specially for functions, to make them available in the different data sheets you can work with at the same time.

Working with data

To be able to work with large datasets and to do things for example like modifying their format to a custom format or automatizing operations related to data entry you will need after activating the macros and opening your workbook or worksheet you will need to declare variables to be able to process the different cells values or formats.

In Visual Basic for Applications (VBA) for Excel, variables are used to store values that can be used in your code. Variables can hold different types of data such as numbers, text, dates, and objects.

In VBA, variables are declared using the "Dim" statement followed by the variable name and the data type. For example, to declare a variable named "myNumber" as an integer, you would use the following code:

Dim myNumber As Integer

After declaring a variable, you can assign a value to it using the equals sign (=) operator. For example, to assign the value 10 to the "myNumber" variable, you would use the following code:

myNumber = 10

You can also declare and assign a value to a variable in a single line of code. For example:

Dim myText As String: myText = "Hello, world!"

In addition to simple data types, VBA also supports object variables, which allow you to manipulate objects in Excel such as worksheets, ranges, and charts. To declare an object variable, you would use the "Set" keyword followed by the object name and the object type. For example:

Dim myRange As Range

Set myRange = Range("A1:B10")

Once a variable has been declared and assigned a value, you can use it in your code to perform various operations. For example, you might use a variable to perform calculations, manipulate text, or update the value of a cell in a worksheet.

Overall, variables in VBA for Excel are an essential tool for storing and manipulating data in your code. By using variables, you can create more flexible and dynamic VBA programs that can perform a wide range of tasks.

Now that you have understood what variables are and a Little bit about coding you can use the insert tab of your embded Excel code editor to create a user form. You will see how a blank grey square appears in the middle of your code editor.

User forms are a key tool in order to manage data, when executed, they place extra options on the screen of the user's worksheet and allows you to automatize data input and many other processes with the use of buttons, text boxes, radio buttons, labels and others.

Now you will see a tolos menu beside the grey square, select one text box and drag it to the grey area, you will see a toggle button aswell, drag one to the grey area.

Now double clic on the toggle button and write this piece of code between the private sub togglebutton1_click() like this:

```
Private Sub ToggleButton1_Click()

MsgBox (TextBox1.Value)


End Sub
```

Then, with the code editor at the front press F5 to launch the macro and you will see a grey customizable window appearing infront of your screen, please then type anything you want in the text box and clic the toggle button, you will see a message window appear on your screen with the same text you typed.

What is a subroutine:

In Visual Basic for Applications (VBA), a subroutine (or sub for short) is a block of code that performs a specific task. Subroutines are used to organize code into logical units and make it easier to read and maintain.

To declare a subroutine in VBA, you use the "Sub" keyword followed by the subroutine name and any parameters you want to pass to the subroutine in parentheses. For example, the following code declares a subroutine named "HelloWorld" that takes no parameters:

```
Sub HelloWorld()
    ' Subroutine code goes here
End Sub
```

You can also declare parameters for your subroutine by specifying their names and data types in the parentheses after the subroutine name. For example, the following code declares a subroutine named "AddNumbers" that takes two integer parameters:

```vba
Sub AddNumbers(num1 As Integer, num2 As Integer)
    ' Subroutine code goes here
End Sub
```

Inside the subroutine, you can use various VBA statements and functions to perform tasks such as manipulating data, displaying messages, and interacting with Excel objects.

To call a subroutine from another part of your VBA code, you simply use the subroutine name followed by any required parameters in parentheses. For example, to call the "HelloWorld" subroutine, you would use the following code:

```vba
HelloWorld()
```

Subroutines can also return values by assigning a value to the subroutine name. For example, the following code declares a function named "GetAverage" that takes an array of integers as a parameter and returns the average of the values in the array:

```vba
Function GetAverage(nums() As Integer) As Double

    Dim total As Double

    Dim i As Integer


    For i = 0 To UBound(nums)

       total = total + nums(i)

    Next i


    GetAverage = total / (UBound(nums) + 1)

End Function
```

Overall, subroutines are a fundamental part of VBA programming, and they allow you to create modular and reusable code that can perform a wide range of tasks. By understanding how to declare and use subroutines in VBA, you can create more powerful and flexible VBA programs that can automate many tasks in Excel.

You can launch your code by pressing on the "play" button that you can find under the format tab in the code editor and stop the execution of the code by pressing the square button that is above the pause button just beside this play button.

In VBA for Excel, there are several data types that can be used to declare variables. Here is a list of the most commonly used variable types in VBA and an example codeo f you to declare them:

1 Boolean - stores a logical value of True or False

Dim myBoolean As Boolean

myBoolean = True

2 Byte - stores a whole number between 0 and 255

Dim myByte As Byte

myByte = 200

3 Integer - stores a whole number between -32,768 and 32,767

Dim myInteger As Integer

myInteger = 10000

4 Long - stores a whole number between -2,147,483,648 and 2,147,483,647

Dim myLong As Long

myLong = 1000000000

5 Single - stores a single-precision floating-point number with a range of approximately -3.4 x 10^38 to 3.4 x 10^38

Dim mySingle As Single

mySingle = 3.14159

6 Double - stores a double-precision floating-point number with a range of approximately -1.8 x 10^308 to 1.8 x 10^308

Dim myDouble As Double

myDouble = 3.14159265358979

7 Currency - stores a currency value with up to 4 decimal places

Dim myCurrency As Currency

myCurrency = 100.50

8 Date - stores a date and time value

Dim myDate As Date

myDate = #3/10/2023 2:30:00 PM#

9 String - stores a sequence of characters

Dim myString As String

myString = "Hello, world!"

10 Object - stores a reference to an object in Excel such as a worksheet, range, or chart

Dim myObject As Object

Set myObject = Worksheets("Sheet1")

Overall, the choice of variable type depends on the specific needs of your VBA program. By understanding the available variable types and their characteristics, you can create more efficient and effective VBA code in Excel.

It is important to declare the variables in your code, if you don't, Excel will automatically asign a default type of variable to the different text operators you type on you code but it can lead to coding errors and misleading results

State aswell that your code is read by the machine from top to bottom and it will stop the execution whenever the code has structural or typographic errors.

In your code editor you can aswell find the properties tab placed to the bottom left of your screen for each one of the elements of your form, there you can setup for example the default values for your text form, his caption (what is show on screen by default)

You might know aswell that this embded code editor accessible by pressing ALT + F11 or from the macros button available in the view tab is based on visual basic for applications and that it's available aswell for other Microsoft office applications such Access and Word, but in this book we will only focus on the main Excel routines that you will need to kick start your carreer as a profficent Excel data manager.

Chapter 6 Excel VBA

In this chapter we will introduce you to Visual basic for applications for Excel and their main practical uses.

We will show you aswell how to créate your own customized Excel functions that if they are stored in the module file they will be available to be used as a regular Excel function as explained before.

In VBA for Excel, a function is a block of code that performs a specific task and returns a value. Functions are similar to subroutines, but they differ in that they always return a value, while subroutines do not have to return a value.

Here is an example of a simple function in VBA:

```
Function AddNumbers(num1 As Integer, num2 As Integer) As Integer
    AddNumbers = num1 + num2
End Function
```

In this example, the function is named "AddNumbers", and it takes two integer parameters named "num1" and "num2". The function calculates the sum of the two numbers and assigns the result to the function name using the "=" operator. The "As Integer" keyword specifies the data type of the value that the function returns.

To use this function in your VBA code, you can simply call it by name and pass in the required parameters, as shown in the following example:

```
Sub TestAddNumbers()

    Dim result As Integer

    result = AddNumbers(3, 5)

    MsgBox "The sum is: " & result
End Sub
```

In this example, the TestAddNumbers subroutine calls the AddNumbers function with the parameters 3 and 5. The function returns the value 8, which is assigned to the "result" variable. The subroutine then displays a message box with the sum of the two numbers.

Here is another example of a more complex function in VBA:

```
Function FindLastRow(ws As Worksheet, colNum As Integer) As Long

    Dim lastRow As Long

    lastRow = ws.Cells(ws.Rows.Count, colNum).End(xlUp).Row

    FindLastRow = lastRow
End Function
```

In this example, the FindLastRow function takes two parameters: a worksheet object named "ws", and an integer column number named "colNum". The function uses the worksheet's Cells property to find the last row of data in the specified column, and assigns the result to the "lastRow" variable. The function then returns the value of "lastRow".

To use this function in your VBA code, you can call it by name and pass in a worksheet object and a column number, as shown in the following example:

```
Sub TestFindLastRow()

    Dim ws As Worksheet

    Dim lastRow As Long


    Set ws = ThisWorkbook.Worksheets("Sheet1")

    lastRow = FindLastRow(ws, 1)


    MsgBox ("The last row is: " & lastRow)

End Sub
```

In this example, the TestFindLastRow subroutine sets the "ws" variable to a worksheet object named "Sheet1" in the current workbook. The subroutine then calls the FindLastRow function with the "ws" object and a column number of 1. The function returns the last row of data in column 1, which is assigned to the "lastRow" variable. The subroutine then displays a message box with the last row number.

Overall, functions are a powerful tool in VBA for Excel that allow you to create reusable code that can perform a wide range of tasks and return useful values. By understanding how to create and use functions in VBA, you can create more efficient and effective VBA programs in Excel.

This code is very similar to VB.net and this book is showing you the specific code to use in Excel, but still many of the methods that we are going to show you are compatible to be used in visual studio.

If you clic on the macros section of the view tab of Excel you will be able to record a macro, which is an automatic and intuitive way to record a simple macro, what this function will do is to record your mouse movements to create and note the correspondent code in the code editor from the actions you perform with your mouse. It is not a very complex way to write code but it will not necessarily produce what you exactly want because it is clearly better to write the code to make sure it will perform the precise actions you require.

Now I will introduce the different and basic code structures for when writing VBA code that should help you develop all the applications you might have in your mind with all your datasets or for educational purposes:

The IF function in VBA is a logical function that evaluates a condition and returns one value if the condition is true, and a different value if the condition is false. The syntax of the IF function in VBA is as follows:

If condition Then

    statement(s) for true condition

ElseIf condition-n Then

    statement(s) for true condition-n

Else

    statement(s) for false condition]

End If

The "condition" is an expression that is evaluated as either true or false, and the "statement(s)" are the code that is executed if the condition is true or false. The optional "ElseIf" and "Else" blocks allow for additional conditions to be tested and additional code to be executed depending on the outcome of those tests.

Here is an example of a simple IF function in VBA:

```vba
Sub TestIf()
    Dim x As Integer
    x = 10

    If x > 5 Then
        MsgBox ("x is greater than 5")
    End If
End Sub
```

In this example, the TestIf subroutine declares an integer variable "x" and assigns it the value 10. The IF statement checks whether "x" is greater than 5, and if it is, it displays a message box with the text "x is greater than 5".

Here is another example of a more complex IF function in VBA:

```vba
Function FindGrade(score As Integer) As String
    Dim grade As String

    If score >= 90 Then
        grade = "A"
    ElseIf score >= 80 Then
        grade = "B"
    ElseIf score >= 70 Then
        grade = "C"
    ElseIf score >= 60 Then
        grade = "D"
    Else
        grade = "F"
    End If

    FindGrade = grade
End Function
```

In this example, the FindGrade function takes an integer parameter "score" and returns a string value representing the letter grade for that score. The IF statement checks the value of "score" and assigns the appropriate letter grade to the "grade" variable. The function then returns the value of "grade".

To use this function in your VBA code, you can call it by name and pass in an integer score, as shown in the following example:

```
Sub TestFindGrade()
    Dim score As Integer
    Dim grade As String

    score = 85
    grade = FindGrade(score)

    MsgBox ("Your grade is: " & grade)
End Sub
```

In this example, the TestFindGrade subroutine sets the "score" variable to 85, and calls the FindGrade function with "score" as the parameter. The function returns the letter grade "B", which is assigned to the "grade" variable. The subroutine then displays a message box with the grade.

Overall, the IF function is a powerful tool in VBA for creating logical statements and making decisions based on conditions. By understanding how to use the IF function in VBA, you can create more efficient and effective VBA programs in Excel.

The Do loop is a type of loop statement in VBA that allows you to repeatedly execute a block of code while a condition is true. The syntax of the Do loop statement is as follows:

Do While condition

   [code statements]

Loop

The "condition" is an expression that is evaluated as either true or false. The code statements within the loop are executed repeatedly as long as the condition is true. The loop will continue until the condition becomes false.

Here is an example of a simple Do loop in VBA:

```
Sub TestDoLoop()
    Dim i As Integer
    i = 1

    Do While i <= 10
        MsgBox i
        i = i + 1
    Loop
End Sub
```

In this example, the TestDoLoop subroutine declares an integer variable "i" and initializes it to 1. The Do loop statement checks whether "i" is less than or equal to 10, and if it is, it displays a message box with the value of "i" and increments "i" by 1. This loop will continue to execute until "i" is greater than 10.

You can exit a Do loop using the Exit Do statement. This statement allows you to exit the loop prematurely when a certain condition is met. Here is an example of a Do loop with an Exit Do statement:

```
Sub TestDoLoopExit()

    Dim i As Integer

    i = 1


    Do While i <= 10

        If i = 5 Then

            Exit Do

        End If

        MsgBox i

        i = i + 1
```

```
    Loop

End Sub
```

In this example, the TestDoLoopExit subroutine declares an integer variable "i" and initializes it to 1. The Do loop statement checks whether "i" is less than or equal to 10, and if it is, it checks whether "i" is equal to 5. If "i" is equal to 5, it exits the loop using the Exit Do statement. If "i" is not equal to 5, it displays a message box with the value of "i" and increments "i" by 1. This loop will exit prematurely when "i" is equal to 5.

Overall, the Do loop statement is a powerful tool in VBA for repeating a block of code while a condition is true. By understanding how to use the Do loop statement in VBA, you can create more efficient and effective VBA programs in Excel.

Chapter 7: Tricks and shortcuts

General keyboard Shortcuts:

Ctrl + A: Select all cells in the current worksheet.

Ctrl + C: Copy the selected cells.

Ctrl + X: Cut the selected cells.

Ctrl + V: Paste the copied or cut cells.

Ctrl + Z: Undo the last action.

Ctrl + Y: Redo the last action.

Ctrl + F: Open the "Find" dialog box to search for specific content.

Ctrl + H: Open the "Replace" dialog box to replace specific content.

Ctrl + S: Save the current workbook.

F12: Open the "Save As" dialog box to save a copy of the current workbook.

Ctrl + O: Open an existing workbook.

Ctrl + N: Create a new workbook.

Ctrl + P: Print the current workbook.

Ctrl + W: Close the current workbook.

Navigation Shortcuts:

Ctrl + Home: Go to the top-left cell of the current worksheet.

Ctrl + End: Go to the last cell of the current worksheet that contains data.

Ctrl + Up arrow: Move up one cell.

Ctrl + Down arrow: Move down one cell.

Ctrl + Left arrow: Move left one cell.

Ctrl + Right arrow: Move right one cell.

Ctrl + Page Up: Move to the previous worksheet.

Ctrl + Page Down: Move to the next worksheet.

Editing Shortcuts:

F2: Edit the selected cell.

Enter: Move down one cell after editing a cell.

Shift + Enter: Move up one cell after editing a cell.

Tab: Move to the right one cell after editing a cell.

Shift + Tab: Move to the left one cell after editing a cell.

Ctrl + D: Copy the contents of the selected cell to the cells below it.

Ctrl + R: Copy the contents of the selected cell to the cells to the right of it.

Ctrl + ;: Insert the current date.

Ctrl + Shift + ;: Insert the current time.

These are just some of the commonly used keyboard shortcuts in Microsoft Excel. There are many more available, and you can even customize your own shortcuts in Excel. To view the full list of shortcuts, you can press the "Alt" key in Excel, which will display a list of keyboard shortcuts for the currently selected tab.